FORUM FOR SOCIAL STUDIES

FSS Studies on Poverty No. 4

Environment, Poverty and Conflict

Papers by
Tesfaye Teklu
and
Tesfaye Tafesse

Addis Ababa
Forum for Social Studies
October 2004

Layout by: Mihret Demissew

Table of Contents

Figures

Tables

Introduction

The two papers published here were presented at the FSS Panel on *Environmental Conflict* at the Second International Conference on the Ethiopian Economy, organized by the Ethiopian Economic Association and held here in Addis Ababa from 3 to 5 June 2004. This was the second panel hosted by FSS at EEA's international conference; the first one was on poverty and was published as FSS Studies on Poverty No. 1. The panel attracted a good sized audience and though there was not enough time to cover all the issues raised by the two papers there was a lively discussion at the end of the presentations.

The term 'environmental conflict' may refer to struggles over natural (or environmental) resources. In this country, environmental conflicts have frequently been caused by competing claims over land, pasture, forests and water, for their intrinsic as well as symbolic value.

Here as elsewhere, the environment, far from being neutral, is eagerly sought by a diverse set of social, economic and political actors and the object of intense competition and conflict among them. Individuals, communities, economic actors, the state, and in recent times multinational corporations have fought over the definition, access and utilization of the environment and its resources. The environment is a vital economic asset, the chief source of livelihood for some and of profits for others. It is also the site where the often antagonistic relation between the dominant and subordinate classes is played out. Moreover, the environment is frequently valued by communities as a symbol of cultural identity, and a source of religious meaning. Thus, environmental arguments are not merely arguments about nature and the landscape, but have complex overtones that are difficult to disentangle from socio-economic, cultural and identity arguments.

We are publishing these papers in this series in the hope that they will stimulate debate and encourage further research on a subject that has not received as much attention as it deserves

Forum for Social Studies
October 2004

1

Natural Resources Scarcity and Rural Conflict:
Case Studies Evidence on Correlates from Ethiopia

Tesfaye Teklu

1. Introduction

There are important *trends* that indicate growing scarcity of natural resources and vulnerability to human impoverishment and scarcity-induced conflicts[1] in rural Ethiopia. First, degradation of natural resources, particularly renewable resources, is widespread as evident from loss of forest, soil and water resources[2]. Second, these resources are increasingly scarce because of diminished supplies, increased population-induced demands, and inequality in distribution[3]. There are signs of growing scarcity of natural resources, particularly arable land in the highlands, and rangeland and fresh water resources in pastoral dominated lowlands. For example, size of operated farmland is declining and fragmented, rural landlessness is on the rise, land rentals on arable and grazing lands are increasing, and access to freely available common resources is diminishing in the highlands.

Third, the rural population is increasingly vulnerable to poverty and food insecurity, particularly in ecologically fragile and marginal agricultural areas. The long-term land productivity is low and, in some cases, it is declining because of declining soil fertility, expansion of the agricultural frontier into marginal areas and low technological change. Declining productivity in the face of land scarcity and absence of alternative employment means the rural population is increasingly vulnerable to impoverishment and hunger. Famine conditions persist due to environmental degradation (i.e., increased aridity and continuous cumulative natural resources degradation), declining agricultural productivity and related income sources, fragility of rural markets under stress, and thinning of private risk coping mechanisms (see, for example, Tesfaye, 2003 on key environment-poverty-famine links and their persistence overtime).

---

[1] *The term conflict here refers to disputes and clashes that arise from claims and command over agricultural land and fresh water resources. The focus here is in particular on environment-induced conflicts between two or more societal groups or communities. Conflict area refers to a physical place where there is potential or actual conflict.*

[2] *Areas with extensive soil and water erosion are marked by low topsoil, declining soil fertility and increased moisture stress and water scarcity. These biophysical changes are associated with scarcity of land and higher rental value, declining productivity, and increased rainfall-linked production variability, impoverishment and out-migration.*

[3] As elaborated in Homer-Dixon (1991, 1994), there are three identifiable processes that cause scarcity of renewable finite natural resources: (i) supply-induced scarcity arising from net decline in quantity and quality of resources (i.e., net decline in stock of natural resources); and (ii) demand-induced scarcity arising from increase in demand for resources (i.e., scarcity arising from demand exceeding supply of resources) and/or unequal access to resources (i.e., change in access and control of resources even when the stock of resources is fixed).

Fourth, there are several cases of environment-induced disputes and conflicts in different parts of the country[4]. Disputes and conflicts arise, for example, over allocation of parental land, returnees claiming ancestral or original land, migrants encroaching indigenous land, and sharing of common resources (i.e., common grazing, water and forest resources). While there is no strong empirical evidence, there is reason to believe that the frequency as well as intensity of conflicts has increased over time.

These trends are not independent processes and, as this paper hypothesizes, natural resource scarcity in a context of high population pressure, poverty, societal heterogeneity, and weak governance and institutions is bound to contribute to dispute and conflict. Yet a knowledge gap exists in terms of our understanding of the extent to which environmental factors have been one of the underlying causes of conflict in Ethiopia. An improved understanding of the causal link between environmental insecurity and violent conflict or at least the correlates will close this knowledge gap for improved conflict management, informed development policy making, and increased likelihood of achieving peace and stability.

The thrust of this exploratory paper is to examine illustrative cases of disputes or conflicts triggered and/or aggravated by scarcity of natural resources, particularly arable land and fresh water resources, and their correlates in the Ethiopian context. It draws on three illustrative cases from the existing pool of conflict cases[5], identifies their correlates, derives analytical framework to demonstrate their links within the African context, and offers concluding remarks related to lessons learned and knowledge gaps.

The paper has five parts. Part one is the introduction with emphasis on trends that are related to rural conflicts. Part two presents the three illustrative cases of rural conflict in Ethiopia. Part three is on rural conflict and correlates. Part four frames the discussion in the context of the African experience. Part five is the conclusion remarks with implications for future research.

2. Three Cases of Rural Conflict

The following three cases represent conflicts between farming communities (farmer-farmer conflicts), farming and pastoral communities (farmer-herder conflicts), and pastoral communities (herder-herder conflicts). Conditions of population pressure, natural resources degradation and impoverishment push highland cultivators to travel long-distance en masse. The first case represents incidences of conflict between indigenous and migrant populations at points of settlement. Conflicts between cultivators and herders are common in the transition zone of the highland and lowland continuum where

---

[4] Asnake (2004) accounts larger sets of conflicts in post-1991 that include conflicts arising from issues of ethnic identity (e.g., Silte-Gurage), control over ethnic territories (e.g., the Borana-Oromia and Gerri-Somali or the Afar and Issa-Somali), and rights of minorities in regional states controlled by an ethnic majority (e.g., the Amhara settlers in Oromia State Region). Competition for access and control of scarce environmental resources underline in some of these conflicts.

[5] Please refer to Tesfaye, Belay and Dessalegn (2003) for the recent review of the literature and details of the known environmental induced case studies (www.paes.org).

2

typically highland cultivators, and agropastoral and pastoral herders converge in a narrow space and compete over scarce resources (such as arable land, pasture and water). The tale of the two villages along the highland and lowland continuum in Northeast Ethiopia is a story of armed conflict between the Amhara (cultivators), Oromo (agropastoral), Argoba (agropastoral) and the Afar (pastoral). There are numerous cases of inter-community conflicts between different pastoral societies living adjacent to each other. The most significant conflict occurs in the Awash River basin, which cuts across different ethnic groups. The third case captures the conflicts in the middle and lower Awash.

i) Resistance to large influx of migrants

Mobility of population is a major mechanism first for easing pressure on land through migration to areas with low population density. Second, migration provides a way to pool climatic risk across space, which is a common practice among pastoral populations in arid and semi-arid environments. Third, migration is a common coping mechanism in time of food crisis. Migration as a way to cope with scarcity of land, seasonal or persistent food shortage, and pool climatic risk has a long history in Ethiopia (e.g., Bahru, 1991 and Daniel, 1990). As the history of long migration shows, people move from densely populated highlands such as the long settled and degraded northern highlands and settle in distant low-density areas.

However, a large influx of migrants at times precipitates armed conflict. For example, waves of migrants from the Wello highlands (Northeast Ethiopia) began settling in the lowlands and escarpments of Northeast Wallaga (Southwest Ethiopia) in the late 1960s (Chernet, 1988). Over the period of the 1970s and 1980s, these migrants established legal rights to the lands they occupied and thereby de-legitimized the claims of the indigenous Oromo people over their ancestral lands. The migrants emerged as the majority dwellers in some of the localities and assumed local administrative power (Chernet, 1988; Assefa, 1999). They also managed to penetrate and settle into the highlands, which were the traditional settlement areas of the Oromo population.

Similarly, hundreds of thousands of impoverished farmers from the famine-affected and environment stressed highlands were settled in the Northwest lowlands (i.e., Mettekel area) in the mid-1980s (Wolde-Selassie, 1997). The settler population was different from the indigenous inhabitants (i.e., the Gumuz) with respect to language and culture. The indigenous inhabitants were dependent on slash and burn farming system, hunting and forest products for their living. The settler population in contrast was dependent on ox-plow cultivation and intensive agriculture.

Often, the indigenous inhabitants resented and resisted the waves of these migrants who moved en masse and occupied ancestral lands (Chernet, 1988; Assefa, 1999; Wolde-Selassie, 1997). First, the indigenous people lost the land of their ancestors. Second, the shrinkage of land diminished the ability of the indigenous inhabitants to practice their traditional farming systems such as descending periodically into the lowlands to practice slash and burn agriculture. In addition, the indigenous people were restricted in their traditional livelihood choices such as extracting value from natural forests. Third, the recipient population perceived that the government land tenure and settlement policies were biased favored the migrant population. Finally, some indigenous people ascribed their worsening living conditions to the presence of migrants and resented the economic ascendancy of some of the migrants. These resentments and grievances grew over the years and were expressed in disputes and physical conflicts with individuals the migrant communities (Assefa, 1999; Wolde-Selassie 1997).

Some of the gains of the migrants that pre-dated 1991 (or, the sources of grievances of the indigenous Oromo) have reversed after the collapse the military regime in 1991. The ethnic and linguistic based policy of the current government entrusts more political and administrative powers to the indigenous population. There is also a shift in military balance in favor of the majority ethnic groups. In an environment of zero sum culture (or, paradigm), the drive is towards excluding the non-indigenous from sources of political and administrative power. Conditions are such that extremist forces are emerging to exploit the existing ethnic cleavages and animosities between indigenous and the migrant populations against the back drop of natural resources scarcity and impoverishment (see Asnake, 2004 and the references their in for recent accounts of rural conflicts).

ii) The tale of two villages in the northeast highlands

Tach Akesta village is found in the 3000 to 3500 masl elevation range along the highland-lowland continuum in South Wello. It has a varied topography, but is mainly hilly and mountainous. Temperature is cool and cold, and the severity of cold temperature in the upper "Dega" restricts the choice of crops grown. Rainfall is low and variable. Drought occurs frequently. Land degradation is severe. Arable land is scarce. There are large numbers of rural households without land. Whilst crop farming is the basic livelihood, farm productivity is low and declining. Farmers are increasingly dependent on income sources mainly from casual wage labor and petty trade.

Balchi Tikuri is located in the lower elevation in the transition from highland to lowland. In the upper catchments are mainly the Amhara settlers practicing intensive crop-livestock farming system. In the lower catchments are largely agro-pastoral Oromo and Argoba, who are shifting to crop farming but still depend heavily heavy on raising livestock. Degradation is not as severe and soils are fertile, but both arable and grazing lands are scarce because of rapid population growth fueled by the influx of recent migrants. Pastureland in particular is in decline because of the expansion of crop cultivation and increasing livestock population albeit shrinking carrying capacity.

The majority of the populations in both villages are poor and food insecure, and their living conditions are worsening. Whilst there is intense competition for scarce land and water in both villages, open armed conflict occurs in the lowland village between distinct ethnically differentiated social groups. The conflicts are mainly between the semi-nomadic Oromo and the nomadic Afar in the lower catchments, but spills to the Amhara settled farmers in the upper catchments. The conflicts between the Oromo and the Afar are mainly over grazing land, but the Afar also claim jurisdictional right over land administered in the Oromo zone.

iii) Conflict over rangeland and fresh water in the Awash Basin

The Awash basin is the home to both the Afar and the Oromo pastoral groups (the Kereyu, the Arsi Oromo and the Ittuu Oromo). The Afar is the largest group inhabiting mainly the middle and lower parts of the basin. The Karrayu are transhumant pastoralists inhabiting the middle valley of the Awash River Basin (see the details in Ayalew, 2001). Typically, these pastoralists raise mixed herd of large animals with small ruminants to take advantage of diversified forage needs and minimize losses due to poor pasture and water conditions. And, because of spatial variation in availability of forage and water, which is seasonally differentiated, the pastoral households move their animals between wet-season and dry-season grazing areas. Such mobility prevents overgrazing and provides equilibrium between grazing and pasture/water resources.

The pastoral territory traditionally inhabited by the pastoral communities is on decline because of changing supply and demand conditions. The supply of pastoral land is declining because of degradation of rangeland, increasing aridity and desertification at the margin, withdrawal of pastoral land for irrigated farming and national parks, and continuous encroachment of neighboring sedentary farmers and non-indigenous pastoralists into the traditional grazing areas. In addition, wealthy and powerful native herders enclose land that is customarily used for common grazing (Getachew, 2001). Consequently, the prime land near the Awash River that was used seasonally by pastoralists during the dry season and during droughts is no longer accessible. The pastoral population is vulnerable to increased impoverishment and famine risk because of shrinkage of pastoral area and range productivity, increased risk of livestock loss, shift to low-return income sources, and degeneration of indigenous support institutions.

Overtime, these changes have disrupted the traditional wet and dry season migration patterns and the equilibrium between environment and population. The pastoral population is increasingly pushed into marginal grazing areas. The Afar clans have moved in a generally westerly direction over the last 50 years (Ali, 1994). They have pushed deeper and deeper into territory used or claimed by the pastoral and farming communities, such as the higher elevations in South Wello in the transition between the highland and lowland continuum. The Karrayu pastoralists in middle Awash, who lost their traditional grazing land, have modified their migratory patterns with significant encroaching onto the traditional territories of the Afar and the Arsi Oromo (Ayalew, 2001). The Arsi herders, who are encroaching onto the traditional territory of the Karrayu in the Awash Basin, are pushing southward in the Lakes region (outside the Awash Basin) and claim land along the border currently inhabited by the Sidama (Bereket, 1999).

The waters of the Awash River and the flood plains along its banks have been the object of intense competition and conflict among a number of interest groups for well over half a century (e.g., Flintan and Imeru 2002; Getachew, 2001; Ayalew, 2001; Irwin, 2000; Seyoum and Yacob, 2000; Ali 1994). The intense competition for resources that has followed has frequently led to serious conflict including violence and armed clashes between on group or another. The Afar clans are continuously in conflict with the Issa Somali and the Ittu Oromo in the east. Conflict also continues to the south with the Karrayu in contested area around the national park. Within the Afar clans, conflicts also occur between the Weima Afar from the north and the Debine Afar in the south close to the Karrayu territory. The Karrayu are in constant conflict with the Afar in the north and the Arsi Oromo in the south, the Issa Somalia and the Ittu Oromo in the east, and the Argoba in the west. The Arsi Oromo also encroach on the Karrayu wet-season grazing and there have been a series of raids and counter-raids among these groups.

3. Rural Conflict and Correlates

The three illustrative cases point to the relation between competition over scarce natural resources (i.e., arable land, pasture and forest) and rural conflict. The degree of relation between resource scarcity and conflict depends on the importance of a particular resource in its socio-economic functions (e.g., contribution to livelihood and welfare) and/or as an instrument for political control. However, scarcity of natural resources alone is not sufficient to explain the occurrence of violent conflict. There are other contributing factors to environmental induced conflicts.

In the cases of the conflicts between the settlers and indigenous inhabitants in northwest and southwest Ethiopia, for example, the encroachment on ancestral land en masse, denial of access to land belonging to the indigenous populations, and contraction of land resources are the chief sources of resentment. But other factors have also contributed over time: importation of unsustainable farming practices, contraction of traditional sources of livelihoods such as forest products, political and economic ascendancy of the migrant population, weak social integration, and perception of biased government policy favoring the migrants (e.g., settlement policy, tenure policy).

The tale of the two villages of northeast Ethiopia illustrates factors contributing to armed conflicts in the transition zone between the highland and lowland continuum. The majority of the populations in both villages are poor and food insecure, and their living conditions are worsening. Whilst there is intense competition for scarce land and water in both villages, open armed conflicts occur in the lowland village. Besides scarcity of pasture and water resources, other factors work in tandem to contribute to conflict in the lowland village. First, there are three ethnically differentiated groups with differing socio-economic and cultural systems (and livelihood systems) competing over scarce land and water resources. Second, there are contested areas with unsettled jurisdictional claims. Third, the legal system is fragmented. An Oromo residing in Oromo zone or an Afar residing in Afar zone is prosecuted for criminal offense in their respective place of residence. Fourth, there is a breakdown in traditional cooperative reciprocal arrangements. For example, the Oromo farmers now cannot move their livestock into the Afar territory during wet farming season, and the Afar are unable to move their herds during the dry season. Fifth, there are perceptions among the competing groups that not only the state-mandated functionaries have been "unfair" as evident in demarcations of administrative boundaries but also ineffective in the prevention and resolution of conflicts. Unlike the Tach Akesta community (upland village) where the ruling party has a strong political and security apparatus, it is not as firm in the armed conflict areas in the transition zone.

The frequency and intensity of armed conflict are pronounced in the pastoral lowlands. Much of the explanation lies in the contraction of pastoral territory and competition for scarce pasture and water resources. There are additional factors contributing to armed conflict: ethnically differentiated groups, contested territorial jurisdiction, erosion of indigenous common property resource management systems, unregulated common grazing land, fragmented judicial system, failed state in prevention and resolution of conflicts, and strong cultural heritage of using force to resolve conflict in some of the communities. Moreover, the long civil war in Ethiopia and the neighboring countries has made matters worse by imposing hardships on many of the pastoral groups, and at the same time allowing a large inflow of arms into their communities.

In short, neither resource scarcity nor impoverishment is sufficient to explain the transition from low-level dispute to armed conflict. Other factors come into play to induce armed conflicts: large migratory population such as pastoral communities or en masse encroachment of land; ambiguity in territorial jurisdiction; contraction of livelihoods and vulnerability to poverty; low institutional response; deficiency in governance and fragmentation of legal systems; and shift in military balance. The likelihood of conflict tends to increase when environmental insecurity induces population mobility, particularly towards heterogeneous communities. Where these migrants dominate economic and political spheres, the recipient communities become aggravated and are liable to engage in conflict[6]. Conflicts are almost certain to occur where a weak state fails to deliver law and order, good governance (transparent and accountable administration), unbiased and fair policy, and institute effective mechanisms to address and resolve grievances and disputes. And the process becomes a vicious cycle as conflict further erodes the effectiveness of political authority.

The drawing of administrative boundaries has become an aggravating factor to environment-induced conflicts. Asnake (2004) characterizes some of the post-1991 conflicts as "environment-cum-ethnic" (note the explicit insertion of ethnic factor), particularly conflicts between settled minorities and regional majorities, and between communities sharing regional boundaries.

4. Framing the Discussion in the African Context

There are different ways of linking the relation between environmental/natural resource scarcity and conflict-aggravating factors. As shown in the flow diagram below (figure 1), there are multiple pathways and factors that contribute to armed conflict. As the African evidence at large and the Ethiopian cases specifically suggest, competition and control over scarce arable land, fresh water resources and non-renewable natural resources such as the war for diamond are either the origin of these pathways or the prime pathway for

---

[6] On the other hand, nurturing of social ties and economic integration neutralize forces that tend towards armed conflict. A case in point is the relatively peaceful existence of the Gojjam migrants from the Amhara land in Oromo society and the conflicts with the Walloye migrants from the Amhara land in Wallaga.

7

armed conflicts (see, for example, Lind and Sturman, 2002; Hussein, 1998; Markakis, 1998; and Percival and Homer-Dixon, 1994 and 1995 for the African evidence at large).

Consistent with the Ethiopian evidence and the African evidence at large, the framework indicates the transition from competition for scarce resources to environmentally induced conflicts occurs under complex sets of conditions: loss of livelihood and impoverishment, deficient tenure arrangement, large mobility of people, societal heterogeneity, and failed governance and policy. Whilst the linkages between environmental insecurity and conflict are not one to one, environment degradation and scarcity is likely to contribute to conflict under sets of conflict-aggravating conditions.

One of the ways through which environmental scarcity contributes to conflict is through setting in motion livelihood crisis in a typical Malthusian paradigm where conditions of rapid population growth, weak technological and institutional innovations, and undeveloped infrastructure leads to decline in productivity, deterioration in livelihoods (or, loss in livelihoods) and deepening poverty. Given that arable land in Ethiopia is the most limiting input for agricultural production and the prime source of subsistence and livelihood, land scarcity sets in motion livelihood crisis, particularly where non-farm income generating option is absent. Impoverishment as a consequence of environmental degradation and scarcity is widespread in Ethiopia and persists particularly in ecologically fragile marginal agricultural areas (see, for example, Tesfaye 2003; Mesfin, 1984; Dessalegn, 1991). Being poor today and the perception of a future threat to livelihoods and survival create condition favorable to conflict.

The other way through which environmental scarcity contributes to armed conflicts is through what Homer-Dixon (1991, 1994) terms "resource capture" and "ecological marginalization". As in the case of Rwanda (Homer-Dixon, 1995) where scarce resources are captured by powerful groups, concentration of scarce resources marginalizes large and weak segments of the population that often opt to migrate either in search of cultivable land regions that are ecologically fragile (i.e., steep upland slopes, areas at risk of desertification, tropical rain forests) or cross into neighboring countries to escape military prosecution and seek physical safety.

Migration plays an important role as a way to cope with scarcity of land or persistent food shortage, and pool climatic risk. In the three Ethiopian cases of conflict, for example, there is an element of population mobility. And, as the evidence from East Africa and the Lakes Region shows, environmental induced out migration in a multi-ethnic environment act as a powerful factor triggering or aggravating armed conflicts (e.g. Ullman, 1983 for Uganda; Homar-Dixon, 1995 and Bigagaza et al, 2002 for Rwanda). The evidence from Uganda, for example, underscores that the major link between environmental insecurity and conflict occurs through environmental induced migration into communities with different ethnic makeup such as the Banyoro and Batooro (indigenous) and the Bakiga (the migrant cultivators) in the southwest and west, the Bahiima (migrant herders) and the Iteso (the indigenous) in the "cattle Corridor" in the south, and the Karamojong (the migrant herders) and the Iteso in northeast.

The effect of ethnicity in the environment-migration link is particularly marked in the history of Burundi and Rwanda. The people in both countries share a similar colonial legacy. Both countries inherited a society fractured along the Hutu-Tutsi ethnic divide. The years after independence have been marked in both countries by political instability and civil wars along ethnic lines. However, ethnic differences are less important in understanding the dynamics of conflict in Rwanda than are elite competition to control scarce environmental resources (Bigagaza et al, 2002). The conflicts in Burundi are not between the Hutu and Tutsi masses, but conflicts between elites from both ethnic groups in their competition for political power. The successive governments in both countries failed to address conditions of rapid population growth, environmental degradation and scarcity, economic deprivation and vulnerability to droughts, social discrimination and injustice, and deepening animosity and conflicts between the Tutsi and the Hutu. Instead the state apparatus provided a vehicle for politicians to control scarce resources and use them to maintain political power and gain popular support along ethnic-based social cleavage.

As land becomes scarce, it is desirable that rights to land are specified and enforced, and mechanisms are in place to manage land related disputes and conflicts. Failed or incomplete institutional response to meet demands for land rights contributes to dispute and conflict. Land related disputes and conflict arise, for example, where there is denial of access to scarce land, legal uncertainty over multiple claims over land and property, insecurity of tenure, and encroaching on protected areas. The most notable African example is the denial of the Tutsi refugees in Rwanda and the Hutu refugees in Burundi to return to their respective countries and claim their ancestral lands (i.e., the right to return to their property).

The Hutu "revolt" of 1959 in Rwanda led to en mass exile of the Tutsi to neighboring countries, particularly to Uganda. Land abandoned by Tutsi was transferred to the Hutu cultivators. Denial of the right of Tutsi refugees to return to Rwanda culminated into organized armed resistance, which gave way to armed resistance and conflicts as in 1963, 1966 and 1973. The history of Burundi is similar except it is the Tutsi that dominated the governments until the 1990s (Prunier, 1995). The Hutu refugees fled massively to neighboring countries in the civil conflict of the 1972. Their abandoned lands were taken over by internal migrants who were largely Tutsis. The failure to respect the right of the refugees to their property provided the power base for armed rebellions. Finding political solution today in both countries is closely linked to addressing the land issues of the refugees.

Armed conflicts are heightened where there is deficit in governance. Deficit in governance manifests in variety of ways such as absence or weak central authority to enforce law and order, control of scarce resources by interest groups, absence of transparent rules of law and enforcement, inadequate institutional and legal framework, and deficiency in capacity (i.e., manpower, finance and broad-based political support). Weak state deficient in governance aggravates environment-induced conflicts. The process is vicious: poor governance begets societal conflict, which in turn erodes effectiveness of state capacity with detrimental consequences such as breakdown of law and order, and loss of political legitimacy to govern.

Although external to internal conflict-aggravating factors, regional political instability and a shift in military balance could contribute to occurrence of conflict. Armed conflicts often spill to neighboring countries in forms of refugee population, flow of arms, and political as well as military alliance between groups across boundaries. For example, the military ascendancy of the Somali groups in Ethiopia such as their expansion into the Afar territory in northeast and the Borana in the south is linked to the change in the geo-political situation in the Horn such as the economic dependency of Ethiopia on Djibouti and hence its reduced leverage to influence the political and military connection between the government of Djibouti and the Somali groups. In addition, the ethnic-based boundaries in Ethiopia have aggravated conflicts over territorial claims as in the case of the Issa Somali who refuse to abide by the ethnic based territorial boundaries and often dispute territorial claims by the Afar.

The Hutu-Tutsi conflicts in Burundi and Rwanda have engulfed the countries in the Lakes Region as in the eastern Congo, which has become the breeding ground for rebel groups and changing alliances against the Tutsi dominated governments in Rwanda and Burundi. For example, the failure of President Mobutu of Congo to restrain the anti-Tutsi rebel groups to attack Rwanda and Burundi contributed to his downfall in 1997. When the successive government under the leadership of Laurent Kabila shifted alliance and turned to assist anti-Tutsi forces against the governments in Rwanda and Burundi, new alliances were formed where the forces from Rwanda and Uganda supported the Rally for Congolese Democracy (RCD) in the 1998 war against the Kabila government, which was received military support from Angola, Namibia, Zimbabwe and Chad.

The framework below does not complete the consequences of conflicts on human lives, degradation of natural resources, and vulnerability to poverty and survival. Breaking the cycle of environmental degradation, poverty and conflicts requires an understanding of key links between conflicts and land degradation, and costs of conflicts including forgone opportunity of economic growth and poverty reduction.

Figure 1: Environment, Poverty and Conflict Links

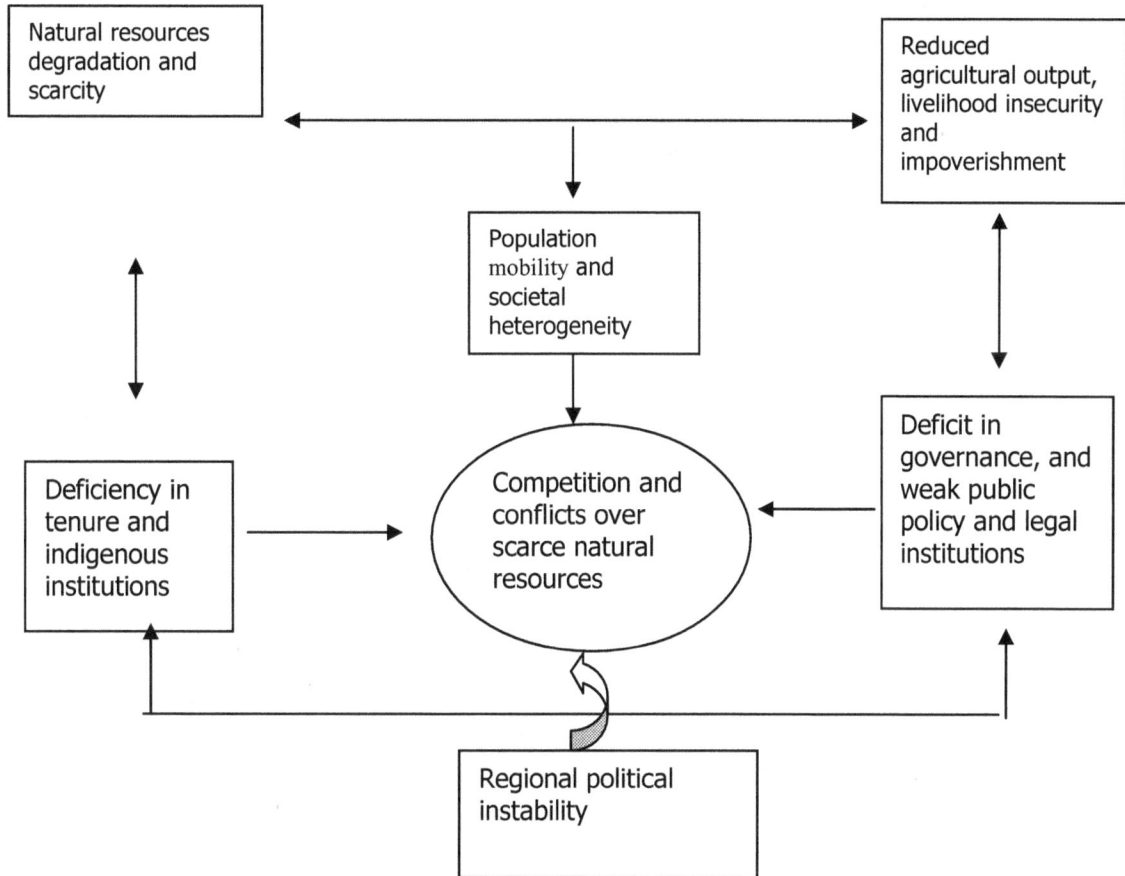

7. Concluding Remarks

There is probably unanimity in the literature that competition for scarce natural resources induces armed conflicts. However, it is not my reading that competition for scarce resources necessarily translates into armed conflict. Conversely, armed conflict does not necessarily occur where there is competition for natural resources. The likelihood of scarcity-induced conflict arises where there are conflict-aggravating conditions: population mobility, social heterogeneity, poor socio-economic integration and heightened vulnerability to poverty and survival, uncertainty in property rights, weak institutions to manage scarce resources and prevent conflicts, and deficient governance and legal framework. Although most of these factors prevail in most of the scarcity-induced conflicts, the sets of factors are location-specific. For example, the factors that explain conflict between the Oromo and the Walloye Amhara in southwest Ethiopia are not exactly the same the conflicts between the Afar and the Somali Issa in the East.

There is a temporal dimension in conflict. For example, incidence of conflicts in drier pastoral areas tends to drop during rainy seasons when bushes and foliages are plenty and when water is plenty. But it peaks in dry season when the resources become scarce. But, more importantly, conflict is a cumulative process. It rarely happens suddenly. Instead it develops over a long period of time. There is a reason to believe that the frequency and as well as the intensity of conflicts have increased over time since the conditions that cause/trigger/aggravate conflicts still persist and deepen such as resource degradation, impoverishment and deficiency in managing conflicts. Ethnic fragmentation is a complicating factor particularly in a context where political and economic elites use it as an instrument to control scarce natural resources.

There are varied and enriching narrations in the literature that indicate there are correlates that work in tandem to explain occurrence of environmental-induced conflict. While our understanding of the correlates is improved, there is still knowledge gap in their direction of causality and relative influence in conflict outcome. It is misleading to focus on a single idea such as environment, tenure or poverty to explain occurrence of conflict when the quantitative evidence is weak to support such claim. For example, it is not questionable that ethnicity plays a role in explaining armed conflicts, but its impact maybe biased where other factors to which ethnicity is a surrogate variable are not controlled. The importance of ethnicity may diminish when the other factors contributing to conflict are controlled in a multivariate regression framework.

In addition to the causality issue, there is also the issue of recognizing that conflict is an outcome and input to environmental change. The emphasis so far has been on conflict as a consequence of environmental change. But conflict has far reaching effect on human lives, and economic and environmental costs. Although calculating these costs is not a simple matter, showing the order of estimates is critical to enhance awareness among stakeholders. It is also important to bear in mind that conflict tomorrow is not the same as conflict today especially where conflict is cumulative and its intensity increases over time. Further research needs to elucidate causal relations between environmental and armed conflicts, assess economic and environmental consequences of armed conflict, and identify pathways for managing conflict.

Preventing conflict is saving lives. It is also conserving the environment and natural resources, and saving scarce development resources to grow out of poverty. Environment induced conflict is preventable. Breaking the cycle of environment and conflict requires addressing environmental concerns and reversing the trends in conflict aggravating factors: livelihood insecurity and impoverishment, defective property rights and governance, and weak institutions managing natural resources and conflict. Public policy has important role in addressing the environment-conflict links. Informed civil societies have a major role in advocating and lobbying policies for addressing environmental concerns within a context of sustainable development and conflict prevention.

8. References

Ali Said (1994), Pastoralism and the State Policies in Mid-Awash Valley: The Case of the Afar, Ethiopia. African Arid Lands, Working Paper Series No. 1/94.

Asnake Kefale (2004), Federalism: Some trends of ethnic conflicts and their management in Ethiopia. In Alfred G. Nhema (ed.), The Quest for Peace in Africa-Transformations, Democracy and Public Policy, International Books, The Netherlands.

Assefa Tolera (1999), Ethnic Integration and Conflict: The Case of Indigenous Oromo and Amahara Settlers in AAROO Addis Alem, Kiramu area, Northeastern Wallaga, Social Anthropology Dissertations Series no.5, Department of Sociology and Social Administration, Addis Ababa University.

Ayalew Gebre (2001), Conflict Management, Resolution and Institutions among the Karrayu and their Neighbors, In Mohammed Salih, Ton Dietz, and Abdel Ghaffar Mohammed Ahmed (eds.), African Pastoralism –Conflict, Institutions and Government, Pluto Press, London.

Bahru Zewdie (1991), A history of modern Ethiopia, 1855-1974. London: James Currey.

Bereket Bezabeh (1999), Ethnic Interaction and Ethnic Conflict: The Sidama and Arssi, B.A. Thesis, Department of Political Science and International Relations, Addis Ababa University, Addis Ababa.

Bigagaza, J., Abond, C., and Mukarubuga, C. (2002), Land Scarcity, Distribution and Conflict in Rwanda, In Lind, J. and Sturman K (2002).

Chernet Wakweya (1988), Land Tenure System and Self-Settled Wolloyes in Abe Dongoro (1900-1974), B.A. Thesis, Department of History, Addis Ababa University, Addis Ababa, Ethiopia.

Daniel Gamachu (1990), Environment and Mass Poverty. In Pausewang, S. et al. (eds.), Ethiopia: Rural Development Options, Zed Books Ltd., London.

Dessalegn Rahmato (1991), Famine and Survival Strategies, The Scandinavian Institute of Africa Studies, Uppsala, Sweden.

Flintan, F. and Tamrat, Imeru (2002), Spilling blood over water? The case of Ethiopia, In J. Lind and K. Sturman (eds.), Scarcity and Surfeit: The Ecology of Africa's conflicts, Institute for Security Studies, Pretoria, South Africa.

Getachew Kassa (2001), Resource conflicts among the Afar of orth-East Ethiopia. In Mohammed Salih, Ton Dietz, and Abdel Ghaffar Mohammed Ahmed, African Pastoralism –conflict, institutions and government. Pluto Press, London.

Percival, V. and Homer-Dixon, T. (1995), Environmental Scarcity and Violent Conflict: The Case of Rwanda. Occasional Paper. Project on Environment, Population and Security, University of Toronto, Canada.

Homer-Dixon, T.F. (1994), Environmental Scarcities and Violent Conflict: Evidence from Cases, International Security, 19 (1).

Homer-Dixon, T.F. (1991), On the threshold: environmental changes as causes of conflict. International Security, 16(2).

Hussein, Karim (1998), Conflict between farmers and herders in the semi-arid Sahel and East Africa: A Review, Pastoral Land Tenure Series No. 10, Overseas Development Group and School of Development Studies. University of East Anglia, Norwich.

Irwin, Ben (2000), Changing Land Use in Pastoral Area, Implications for Increasing Levels of Conflict and Competition between Multiple Resource Users and Multiple

Resource Uses. In Pastorlism and Agro-Pastoralism: Which way forward? Proceedings of the 8th Annual Conference of Ethiopian Society of Animal Production (ESAP). Addis Ababa.

Lind, J. and K. Sturman (eds.), Scarcity and Surfeit: The Ecology of Africa's conflicts. The African Center for Technology Studies, Nairobi and Institute for Security Studies, Pretoria, South Africa.

Markakis, John (1998), Resource conflict in the Horn of Africa. London: The Cromwell Press Ltd.

Mesfin Woldemariam (1984), Rural Vulnerability to Famine in Ethiopia 1958 – 1977, Vikas Publisher, New Delhi.

Ohlsson, Leif (2000). Livelihood Conflicts: Linking poverty and environment as causes of conflict. Swedish International Development Cooperation Agency Department for Natural Resources and the Environment.

Prunier G. (1995), The Rwanda Crisis: History of a Genocide, C. Hurst and Co. Ltd., London.

Seyoum Gebre Selassie and Yacob Arsano (2000), Resource Scarcity and Conflict Management in the Horn of Africa: The Case of North Shoa, Ethiopia, Unpublished paper, Addis Ababa, April.

Teklu. 2003. Environment stress and increased vulnerability to impoverishment and survival: A synthesis, paper presented at the International Symposium on Contemporary Development Issues in Ethiopia, July 11-12, 2003, Addis Ababa and published in the Bulletin of the Forum for Social Studies, 3(1).

Tesfaye Teklu, Belay Tegene, and Dessalegn Rahmato (2003), Environmental insecurity, poverty and conflict in Ethiopia: Aggregate and case studies evidence. Paper submitted to PAES, Uganda (www. Paes.org).

Ullman, R.H. (1983), Redefining Security, International Security 8: 129-153.

Wolde-Selassie Abbute (1997), The Dynamics of Socio-Economic Differentiation and Change in the Beles (Pawe) Resettlement Area, M.A. Thesis, Addis Ababa University, Addis Ababa.

2

The Migration, Environment, Conflict Nexus:
A Case Study from Eastern Wollega

Tesfaye Tafesse

1. Introduction

The northern parts Ethiopia that have been inhabited for a long time, most particularly the highlands, frequently face problems of soil erosion, environmental degradation, land and water scarcity, soil fertility decline, and high population pressure. On top of these severe problems, they are vulnerable to climatic oscillations, mainly to erratic and unpredictable rainfall. These problems in turn expose the regions to recurrent drought and famine. Peasants' responses and coping strategies to these natural as well as human-induced problems are also varied. They include, among others, migration, (voluntary or involuntary)[7] to areas that have a relatively better carrying capacity. Depending upon the severity of the problems, migration could either be seasonal, i.e. short-cycle circulation ranging between one and six months, or permanent. Peasants undertake seasonal migration as a coping strategy to temporarily reduce their household size (the number of mouths to feed) as well as to earn and remit money from the income they obtain from farm and off-farm activities in the recipient areas. In fact, in this type of seasonal migration, which is the norm rather than the exception in rural Ethiopia, the migrants return home at the peak of farming activities. This type of migration actually takes place in most parts of Ethiopia simply because peasants will incur a high opportunity cost if they abandon their farmlands.

On the other hand, peasants normally undertake forced migration as a last resort strategy when all other coping strategies fail. As noted by Corbett (1988), people who live in conditions that put their main source of income at recurrent risk, for example, peasants living in drought-prone areas, will develop self-insurance strategies to minimize risks to their food security and livelihoods. Surke also notes that "for rural people, migration is one of several coping strategies to deal with poverty which in itself reflects a combination of social, economic and political conditions" (1994:86). These may involve, among others, rural-rural as well as rural-urban migrations in search of employment and livelihood in distant farmlands and/or labor markets. One should not, however, overlook the role of kin and peer pressure and the quest for a relatively better life as factors that tempt people to leave their homes for good. It is basically this aspect, viz. spontaneous or involuntary migration and its consequences, i.e. conflict and displacement that is the focus of this paper.

---

[7] According to Gebre (2002), the distinction between voluntary and involuntary migration is not always easy to make because of varying contexts and relationships.

Coping strategies, including spontaneous migration, by specific social groups of the peasantry and pastoralists in Ethiopia can be understood if and only if, besides the development history of the country in general and the specific local areas in particular, one tries to see the interaction processes with external factors and actors. It goes without saying that the capacity to control/overcome crises (famine/hunger) in Ethiopia becomes difficult because the endogenous capacity to get access to food supplies declines. Moreover, the problem becomes complex and insurmountable due to the mismatch between the local problems and needs at the grass-root level and external actors' emergency intervention.

This study will attempt to identify the characteristics of the migrants from the various zones of the Amhara National Regional State (hereafter ANRS) of Ethiopia, who settled in various Peasant Administrations (PAs) in Gida Kiremu Woreda of Oromia National Regional State (hereafter ONRS) and the consequences that followed, viz. violent conflict and displacement. By so doing, attempts will be made to explore the social, economic, political and ecological factors that have forced the migrants to leave their homes for good, the ways and means by which they lived in the host region, and the factors that triggered conflict and displacement. The research will trace the movement of the people migrants-cum-settlers-cum-displacees, and document what has happened in the process. Based on these findings, attempts will be made to raise policy implications in the context of Conflict Prevention, Management and Resolution (hereafter CPMR) mechanisms that could possibly reduce the recurrence of violent conflict in the same region and elsewhere in Ethiopia.

The fact that little or no empirical research has been carried out in the area on such a crucial issue makes the topic pertinent and timely. Besides, the migration-conflict-displacement nexus that will be established by probing into vulnerability, entitlement, and coping approaches will help to give a better picture of the problems that bedevil Ethiopia.

2. Objectives of the Study

- Identify the factors that led the migrants to spontaneously leave their ancestral homes for good (push factors) and to move/settle in Gida Kiremu Woreda (pull factors)
- Analyse the adaptation of the migrants after they settled in Gida Kiremu Woreda – *how the migrants obtained land, how they lived there and the kind of social interactions they had had with the local population*
- Uncover the underlying causes of the violent conflict that led to population displacement
- Suggest plausible CPMR mechanisms to mitigate, and if possible stop, the flaring up of other similar conflicts in the region and elsewhere in Ethiopia.

3. Data Sources and Methodology

It should be stated at the outset that the study is based on primary data that was generated by conducting preliminary household surveys, focus group discussions, and interviews with all the stakeholders in both Regions. In order to get balanced information and possibly establish 'the truth' in and around the conflict situation, the researcher held discussions with the local Oromo population and the Amharas still living in Gida Kiremu Woreda as well as the internally displaced persons' (IDPs) that are presently living in Jawi/Metekel resettlement site. Interviews have also been conducted with authorities in Gida Kiremu Woreda, various ANRS offices, and in Awi Zone where the peasants have been resettled.

Secondary data on various problems related to the study area have also been obtained from Woreda/Zonal offices, ANRS's Regional Disaster Preparedness and Prevention Bureau (DPPB) office, strategy papers, evaluations and studies from NGOs', bilateral organizations and the different tiers of the government.

4. The Study Area

The study area encompasses the migration paths that were followed by Amhara peasants from the various Woredas/Zones of the Amhara Regional State and its environs into Gida Kiremu Woreda, Eastern Wollega, the settlement and eventual displacement of the migrant-cum-settlers, and their subsequent resettlement in Jawi Woreda, Awi Zone (for the location of the areas, refer to Figures 1and 2). Issues related to the resettlement of the IDPs' in Jawi/Metekel (Awi Zone) have been taken out from the study for two reasons (a) the displacees-cum-resettlers still live on emergency food aid that is supplied by the regional DPPC and (b) given the young age of the resettlement site (only three years old), it will be too early to make conclusive judgements.

In the past three decades or so, due to a host of environmental and non-environmental factors such as land scarcity, environmental degradation, population pressure, entitlement decline, political economy/social differentiation, vulnerability, food insecurity, peer and kin pressure and other related factors, thousands of people have spontaneously migrated, at different times[8], from different zones of the ANRS into Gida Kiremu Woreda, East Wollega Zone. Their lives were disrupted when resource-cum-ethnic conflict flared up in the area at the end of 2000 resulting in the displacement of more than 12,000 people (for details refer to preliminary results section). They have now been resettled in Jawi/Metekel resettlement site in Awi Zone.

---

[8] The first waves of migrants who trekked to Gida Kiremu Woreda may be traced back to the imperial times in the late 1960s and early 1970s and the early years of the Derg period, while the most recent ones are as recent as 1999. During the 1980s, another wave of migrants came to the area through the government's resettlement program.

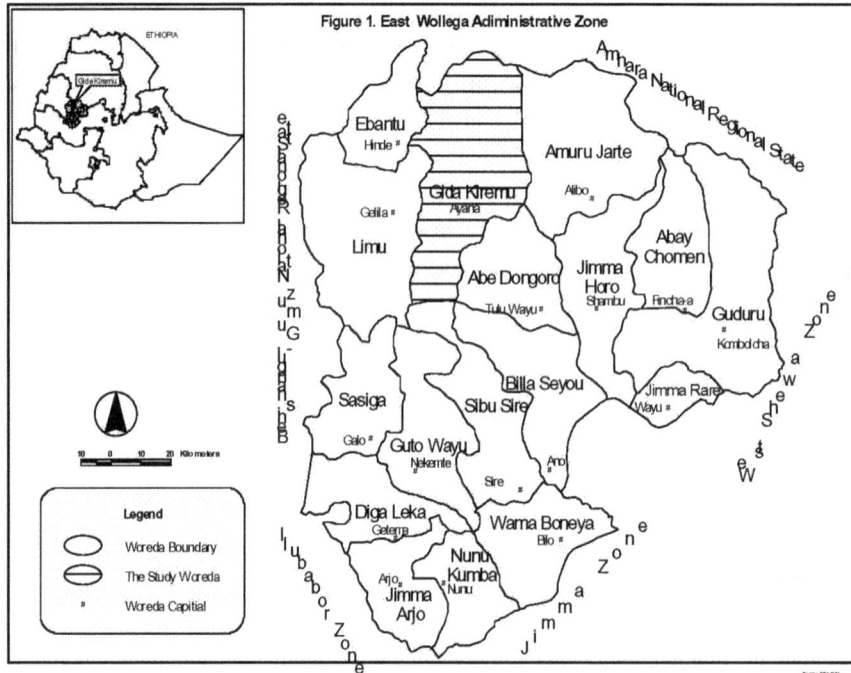

Figure 1. East Wollega Adiministrative Zone

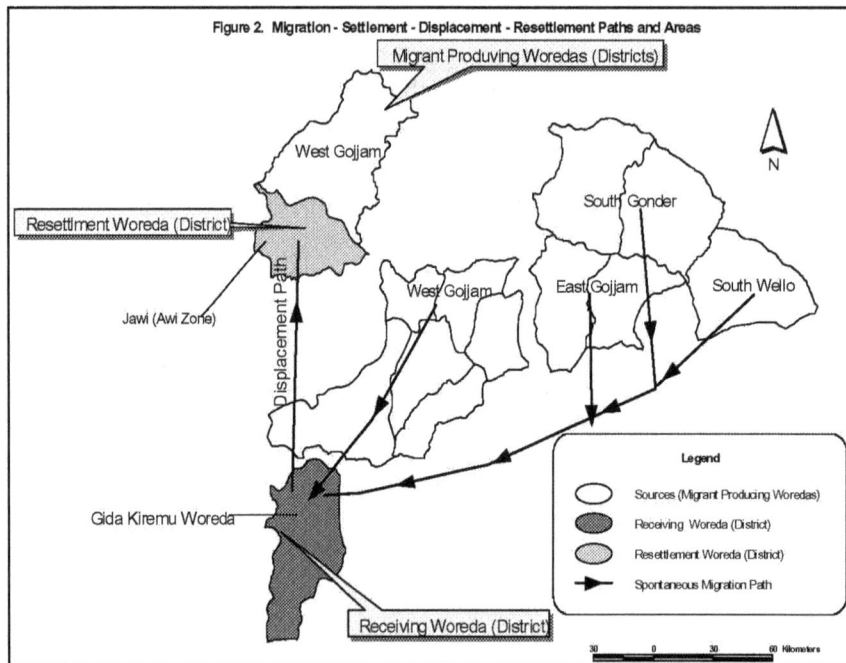

Figure 2. Migration - Settlement - Displacement - Resettlement Paths and Areas

18

5. A Review of the Debate

The causal factors that lead to migration are so complex that it becomes difficult to isolate one factor, say environmental degradation or population growth, as the sole cause. For instance, it will be stereotypical and unrealistic to consider that population growth alone causes people to migrate. In order to examine the causes of migration and its far-reaching consequences, one needs to bring in modern risk and vulnerability research concepts into the limelight to utilize them as a theoretical basis and background. The exposure to stress, shocks, and risks of the poorer segments of the society, i.e. vulnerability, which is normally seen in view of or in relation to the ecological and economic deterioration of the group in vulnerability studies, could be of immense help in unraveling the underlying causes of migration.

Previous assumptions have changed when the understanding of natural catastrophes (e.g. floods, droughts) was limited to the spatial exposition of an area against natural risks/hazards, predicting the probability of their recurrences in some future date and/or modifying their impact through technical interventions. Such an approach that gave emphasis to the violent forces of nature failed to consider the socio-economic dimensions, including the socio-cultural, political-economic, and entitlement contexts. Even the socially differentiated effects of disasters have been ignored, i.e. why some people starve while others do not. This outdated approach had a technocratic overture and was dominated by a natural science-cum-physical geography-based analysis. As stated by Hewitt (1983), in the late 1970s' and 80s', the "human ecology' approach, which emphasized the relationship between population size and resources, was strongly criticized in risk and vulnerability studies for not considering the socio-economic system. Even the socially differentiated effects of the disasters have been ignored.

It is now clearly recognized that there are actually no 'natural' disasters or the naturalness of disaster events as such but rather the ones that have a human dimension that can legitimately be referred to as 'social disasters'. The latter arise as a result of the prevailing social structure and social protection systems that manifest themselves, among others, through unequal access to opportunities and unequal exposure to risks. Inequalities in risk and opportunities are largely a function of power operating in all societies, which are normally analyzed in terms of class, gender and ethnicity. In sum, one can say that disasters, which may resemble those usually blamed on nature, are inherently induced by human action (Cannon, 1994). It serves some political interests to maintain the notion that disasters are natural rather than caused by political and economic processes. If people can be made less vulnerable or non-vulnerable by their situation within the given political and economic systems, then a hazard may still occur, but need not necessarily produce a disaster. So, the major determinants that make people vulnerable are the social, economic and political factors, which in turn determine the level of resilience of people's livelihoods, and their ability to withstand and prepare against hazards.

Vulnerability studies should incorporate the production and social systems as well as the traditional security systems and coping strategies of the vulnerable groups together. That is why many researchers agree on the notion and outcomes of social vulnerability research, which basically focus on socially differentiated vulnerability to crisis and the capacity to cope. Such notions are gradually changing from the old geographical paradigm or discourse that emphasized risk or hazard research towards more conceptual/analytical frameworks that accentuate the human dimension. It should at this juncture be stated that this approach has also geared to another extreme by giving an undue weight to social attributes by sidelining the physical factors. The researcher of this paper would like to use an amalgam approach by which both factors could be entertained. The idea behind this approach rests on the premise that when social, economic and political processes place certain groups of people in a vulnerable situation, natural hazards, such as of drought, can easily trigger disaster.

The Indian Nobel Prize Winner, Amartya Sen, first developed the entitlement approach in the context of famine-crisis research in 1981. Prior to that, in the 1970s', there was a theory based on the food-supply side to explain famine crisis. The analysis of exchange entitlements by Sen (such as selling labor to earn money with which to buy consumption goods) has restored the much-needed emphasis to the role of demand failure, as well as identifying the relative significance of supply-side changes. Entitlement failures, which arise from a mix of socio-economic, environmental and political factors, may lead to the vulnerability of the particular social group resulting in famine and involuntary migration. Corbett (1988) registered her reservations by stating that Sen's analysis has failed to solve the issue of what households decide what to do, given a certain set of entitlements. She further that it seems unlikely that even in the face of declining entitlements, households passively await starvation.

Further developments on the idea of vulnerability were made through the work of Chambers (1989). He differentiated external and internal dimensions of vulnerability whereby the crisis depends upon the exposure against known or defined risk exposures as well as the socially differentiated capabilities to withstand the crisis, namely coping. It should again at this point be reiterated that vulnerability is not necessarily an outcome of natural environmental calamities but rather an outcome of the interrelationship between the existing social framework and political conditions (Mesfin, 1984).

Following Chamber's work, important conceptual developments on the subject of vulnerability have come into the foreground with the works of Bohle and Watts (1993), Bohle et al (1994) and Blaikie et al. (1994). Of these, Bohle et al (1994) combined the traditional 'human ecology' approach with political economy (the manner in which surplus is generated and allocated, social power and control, debt etc.) and entitlement theories to explain vulnerability. This was another landmark achievement that helped in explaining the external dimensions of vulnerability. Contrary to this approach, Blaikie et al (1994) came up with the 'Pressure and Release' (PAR) model in which they designated disaster as the intersection of two opposing forces, namely those processes generating vulnerability on one side and physical exposure to hazards on the other. They argued that an increasing pressure on people arises from either side – from their vulnerability and from the impact and severity of the hazard on those people at different degrees of vulnerability. The 'release' idea is incorporated to conceptualize the reduction of disaster, i.e. to relieve the pressure vulnerability has to be reduced. By so doing, they have identified different forms of causal factors for vulnerability, which when taken collectively can give a holistic picture.

Against the above-stated backdrop, the British development agency, viz. the Department for International Development (DfID), and UNDP developed a practically oriented sustainable livelihood approach (Carney, 1998). They came up with a differential possession of assets over which varied social groups have different controls. This in a way helped in indicating how the groups react and define susceptibility to catastrophes in a vulnerability context or situation. This approach, which gives emphasis to endogenous factors, can fill the gap in existing vulnerability studies that have placed more emphasized on exogenous factors.

One can hence deduce from the above arguments that a host of factors, including but not restricted to socio-cultural, ethnic, agro-ecological, political economic, resource scarcity and conflict, peer and kin factors, social network, population pressure, and processes of social differentiation act in unison to cause migration that often result in violent conflict and population displacement (Wenzel, 2002). It means, for any spontaneous migration to take place multiple, instead of single, causal factors are responsible. One has also to differentiate between short-term (reversible) and long-term (irreversible) causes as well as endogenous and exogenous factors that can cause migration, violent conflict and population displacement. On the one side of the scale, we have factors like climate change acting as exogenous factors and on the other we have environmental degradation or ecological disturbance, the depletion of the resource base, social structures, social network disruptions, food insecurity, population growth and the resultant diminishing of holdings, government policies (e.g. distorted economic policy), political power and entitlements acting both as endogenous and exogenous factors (Ibid.). It can hence be deduced that "movement takes place in response to a combination of environmental, social and political (including armed conflicts) stimuli" (Lonergan, 1998 quoted in Wenzel, 2002).

Besides, there are long and short-term causes that encourage people to migrate with ecological factors coming into the fore as one out of a host of factors. It depends on societal/contextual conditions whether after onslaughts of drought flooding or environmental degradation people migrate or not. This is so simply because a bundle of factors play a collective role in inducing migration. Hence, one has to view migration not as a 'disaster-go' or 'drought-go' phenomenon but rather as a process that takes a longer time and unreserved patience to arrive at. The threshold value or the cut-off point for people to reach at a decision to move for good differs from individual to individual due to the variations in migration behaviors, coping capacities and social networks.

Corbett's idea (1988) which viewed migration as a last in a sequence of household's responses to famine conditions and a clear indication that many other responses have failed, should be taken note of for further empirical investigation. As this author's observation in the study area testifies peasants often use various coping strategies in times of stress, such as borrowing grain, reducing consumption, changing dietary habits, receiving emergency food handouts and yet making a seasonal migration to rural or urban areas at the same time or sequentially as part of the households coping strategy. They have, what Dessalegn called 'anticipatory and crisis survival strategies', "the former being those adopted during periods of normalcy, and the latter in times of stress" (1991:16).

Migration is a result of the prevalence of poverty and famine but also war and conflict, as cases from Ethiopia and Mozambique during the 1980s' have clearly shown. Wars and conflicts disrupt the food and labor markets so much so that households may be unable to obtain sufficient access to food through these channels to supplement their own production deficits. This, in turn, forces people to migrate to other areas. This situation contrasts with Sen's (1981) study of famine in Bangladesh in 1974 where the breakdown of markets was not a contributing factor but that landless laborers were squeezed out of the market by a dramatic decline in their exchange entitlements (Corbett, 1988).

If one needs to identify the factors that initiate migration and the degree to which they are inter-related, one must undertake a study at the local or community level. This would help us to identify the concrete social, economic, ecological and political contexts that generate migration. So, the nexus between increasing food insecurity, which comes as a result of drought and other natural calamities, vulnerability and migration processes is what this study tries to investigate.

Internal conflicts, which do have different causes and take different shapes in different countries, often lead to population displacement. In most cases, as is also the case in some parts of Ethiopia, ethnic differences may be the major cause of conflicts but in reality a host of factors, including but not restricted to resource scarcity, poverty, ethnicity, religion, lack of good governance could act as causes, either individually or collectively. Of these, a lack of good governance could be spelled out as a major cause of conflict in Africa where "unitary systems make the largest group, dominant with a high degree of power centralization where ones interest will be protected at the expense of the minorities" (Adedeje 1999: 42-45). So, in addition to power sharing, ethnic rivalry or animosity and unequal access to resources should be given due emphasis in the analysis of internal political dynamics in Ethiopia in particular and the Horn Region in general.

One of the reasons for an inter-ethnic boundary dispute in many African states is the pattern of settlements that impinge upon shared resources. According to Hutchison, "...borders artificially alter the values of resources, depending on which side of the line they rest" (1991:21). As a result, those groups of people that will be denied the right of access to shared resources can easily trigger inter-ethnic conflicts. The situation becomes worse when natural calamities, such as drought, overwhelm the region at large inducing resource scarcity.

Although much of the intra-state conflicts in the Horn of Africa have the tone of politicized ethnicity, peoples' categorization into majorities or minorities coexisting in a single state can also trigger conflicts. It goes without saying that not all nationalities that inhabit a single state are numerically equal. Problems emerge when inequalities in the size of population are translated into inequalities related to power and resources. To come to terms with the diversity that exists in a multi-ethnic society there should be tolerance towards multiculturalism. As aptly put by Maybury-Lewis, "a state can only function as a multi-ethnic system if its citizens are educated in tolerance and its civil society is working reasonably well" (1997:153). If we take the same author's words a qualifying ethnicity or "not simply [as] an innate propensity of human beings to bond with those like them and to fight with those unlike them" (Ibid, 157), then to what factors can we attribute conflicts or wars that emerge as a result of ethnicity? Adedeje argues that "apparently ethnic differences may seem to be major causes of conflicts but in reality it is the lack of satisfactory political arrangements and transparency that can act as source of conflicts" (1999:42).

In Ethiopia and some other African countries, ethnicity has been politicized to the extent of inciting bloody ethnic conflicts. Markakis also claims that "there is no doubt that ethnicity, political mobilization on the basis of collective identity based on cultural affinity is frequently though not always, one variable involved in the chemistry of the conflict" (1998:103). The increasing domination and the unwillingness of some African governments to accommodate the interests of other groups lead to conflicts, and to make things worse, there has always been more emphasis on differences among ethnic groups than their similarities. The solution to such a problem rests more on enhancing democratic values and institutions that could facilitate tolerance and magnanimity instead of differences. It would, however, be naïve to consider that democratic orientation and 'modern' political parties could replace traditional tribal and/or ethnic loyalty in Africa. A due recognition of the latter in an African political discourse is a must if we opt to build tolerance and co-existence.

Access to the use of scarce resources is based on the communities' settlement in a specific area and the claims made by those groups to defend it as their own right. The claims may further be intensified if, at times, political tensions or natural calamities (e.g. drought) occur in one geographical area that may force the community to adopt various coping mechanisms in the face of stress, including migration to other places. An outstanding example that happened in Kenya in 1993 that was noted by Markakis (1998) showed how the Kikuyu peasants moved into the Rift Valley in large numbers in the 1960s and how they came under violent attack by Kalenjin tribesmen who claimed prior rights of possession to the region's resources. The Kikuyu migration took place under the Kenyatta regime that was dominated by Kikuyu kinsmen, while the Kalenjin attack occurred when Arap Moi, an ethnic kinsman, came to power in Kenya. As stated by the same author, "although the importance of the land factor was generally recognized, the clashes were widely attributed to tribal enmity inflamed by political passion" (Ibid:4-5). When such conflicts take an 'ethnic shape', it would not be that easy to apply either traditional or 'modern' conflict resolution mechanisms.

6. Preliminary Findings

6.1 Causes of Migration

As has been indicated earlier, in the past three decades or so, due to a host of factors such as land scarcity, environmental degradation, population pressure, entitlement decline, political economy/social differentiation, vulnerability, food insecurity, peer and kin pressure and other related factors, thousands of people have spontaneously migrated, at different times, from different zones of the ANRS into Gida Kiremu Woreda, East Wollega Zone (for the location of the areas under question refer to Figures 1 and 2 under section 4 and for the figures Table 1 overleaf) .

After their arrival in the Woreda, most of the migrants first served as share-croppers with the local population or early Amhara settlers, who in most cases were their own relatives, for a couple of years or so before obtaining their own plots of land for farming.

Table 1. Sources and Destination of Migrants-cum-Displaced Persons

Sources of Migrants (original home)			Recipient Areas of Migrants					Resettlement Sites (after displacement)				
Zone	Woreda	Pop. Size	Zone	Woreda	Peasant Adm. (PA)	No.of HHs	Population Size	Zone	Woreda	PA	# of Households	Population
South Gondar	Semada	6225	East Wollega	Gida	Mirga Giregna	1743	5008			(Mirga Jiregna I) Aima Gabriel	773	2669
	Esete	2740	East Wollega	Gida	Wasti	643	1902			(Mirga Jiregna II) Addisnuro Gabriel	774	2668
East Gojjam	Goncha	1926	East Wollega	Gida	Haro Addisalem	703	1989			(Haroo Addisalem) Addisalem Abo	576	1893
	Motta	239	East Wollega	Gida	Jiregna	366	1151			(Waste) Addismaba Abo	525	1840
	Degadamot	90	East Wollega	Gida	Haro Misema	395	1099	Awi	Dangla	(Sombo) Assefameda Abo	350	1143
	Dembecha	62							(Jawi Resettle ment Site)	(Jiregna) Zurla Woin	332	1320
West Gojjam	Quarit	37								(Haroo Misema) Agere Selam	270	1030
	Jabbi	100										
	Sekela	40	East Wollega	Gida	Sombo	364	1047					
	Burie	8										
	Achefer	4										
	Mecha	3										
Awi	Gimja bet	35										
South Wello	Saint Ajbar	687										
Total		12,196				4214	12,196				3600 *	12263

Source: DPPC (2002). Jawi Resettlement Project Report, Bahir Dar; * The reduction in the number of households is attributed to the return of some of the displaced people to their original abodes and to wollega.

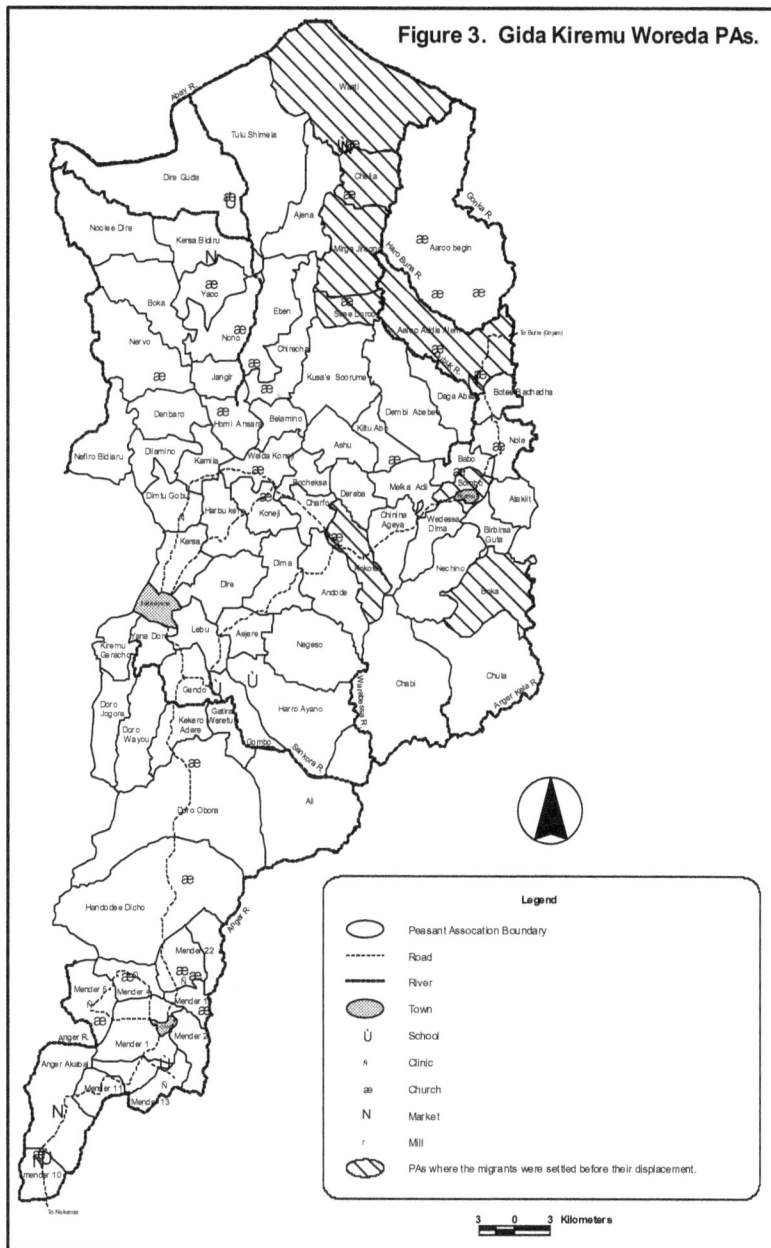

Figure 3. Gida Kiremu Woreda PAs.

Source:- CSA

The plots in most were obtained either through contractual arrangements and/or purchase from the local population. After the settlers obtained land in one of the two ways, they cleared the forested areas and converted them into farmplots. Gradually, they settled themselves in the nine Peasant Administrative areas (PAs) of the Woreda, namely Aaroo Addisalem, Aaroo Bagin, Boka, Chelia, Kofkofe, Kusaye, Mirga Jiregna/Sire Doroo, Sombo, Wasti, as well as in Kiremu town (for the location of the PAs refer to Figure 3). According to the information obtained from the migrants, almost all of them fared better in their new homes compared to their birthplaces: they were able to minimizing their vulnerability and enhance their food security status.

The migrants and officials from both regions have identified a host of factors that persuaded the former to leave their homes for good (push factors) and to choose Gida Kiremu Woreda as their destination (pull factors). According to the informants, the former include, among others, land scarcity, overcrowding, recurrent drought, land and soil degradation, declining fertility, diminishing holdings, inability to pay taxes, inability to sustain the family, fleeing court cases for crimes committed back at home, and the land redistribution that diminished holdings in Gojam. The pull factors, as stated by Zelalem (2003), include factors that could be categorized in three temporal dimensions: (i) *Imperial times*: the local Oromo *balabats* attracted Amhara settlers in order to have a larger base for candidacy in the Imperial Parliament and to expand their tax base. That is how migration into Gida Kiremu Woreda started. (ii) *Derg times*: (a) the 1975 land reform that dispossessed the feudal lords left lots of vacant land that attracted migrants, (b) the settlement pattern in Eastern Wollega where the local population lived uphill in the highlands, left vacant lands in the lowlands which attracted potential migrants (c) the presence of vast, fertile and uninhabited tracts of land in the region. (iii) *post-1995*: the opening up of the Nekemt-Bure road in the early 1990s.

Ato Dejene Getahun, Deputy Administrator of Gida Kiremu Woreda, categorized the migrants into four groups: (a) those who came during the last years of the imperial regime; (b) those who had been resettled in the Woreda after the the 1984/85 drought [note the location of various villages in Figure 3]; (c) those who trekked into the Woreda after the fall of the Derg in 1991; and (d) those who migrated from Gojam, most particularly from Quarit Woreda, after they lost some of their plots due to the land redistribution that took place in 1997/98. According to the deputy administrator, the early migrants and resettlers that fall in the first two categories (a & b) had had no problem in integrating themselves with the local population. He said that some of the migrants married with local inhabitants. Over and above that, stated the same informant, they respect government policies, paid taxes on time, serve in the militia and peasant administrations and registered their firearms with the concerned office. Conversely, the official considers those who migrated to the Woreda after 1991 as 'outlaws' for three reasons: (a) they have illegally seized firearms and refused to get them registered, (b) they have deforested the area in a 'merciless manner', and (c) they evaded registration in PA offices. As a result, stated the same informant, they remained hostile to the local population and the administration alike. The migrants, however, reject most of the accusations that were laid against them as groundless (see details below).

In addition to the above-stated factors, the migrants also pinpointed peer and kin pressure, availability of virgin and fertile land, and suitability of the climatic condition in the destination area as significant pull factors that tempted them to migrate to Wollega. Of these factors, the first two played an important role in the decision of the migrants to move to Gida Kiremu Woreda. Almost all the sampled migrants had had one or more members of their nuclear and/or extended family in Wollega prior to their migration. The migration path normally followed the following course: firstly, male household heads migrated into Gida Kiremu Woreda with their cattle, worked for a year or two as share-croppers, obtained land for farming by buying it or through lease agreements with the local population, cleared forested areas and turned it into farmland. In due course, they would first bring their immediate family members to the area in question to be followed by other family members and close friends. Of the ten randomly selected migrants from Dega Damot, Goncha, Motta, Saint Ajbar, and Semada Awrajas who arrived in Gida Kiremu Woreda between 1984 and 1997[9], all of them had relatives (brothers, father-in-law etc.) who had settled earlier in various PAs of the Woreda. This is ascertained by Assefa who stated that, "...most of my [Amhara] informants reiterated the fact that they were informed about the availability of land in Wollega by those who had already settled in Gida Kiremu Woreda" (1999:26).

As noted above, the migrants and authorities alike have more or less identified dominantly physical/environmental factors that persuaded the migrants to migrate. Although the peasants failed to indicate social differentiation as a cause of migration, they acknowledged its existence stating that there were sharp differences in land size and quality as well as livestock possession in their ancestral homes. They have documented the status of some of the people in their communities who had been more resilient and less vulnerable to dire situations compared to them implying, among others, the existence of unequal exposure to risks. These groups of people included those that had a relatively larger land holding, better quality plots and a sizeable number of livestock. They also explained their decisions to migrate because of the drastic deterioration of their lives from year to year due to declining soil fertility, diminishing holdings and recurrent drought. If one adds to these the prevalence of weak traditional protection systems that hardly shielded them from the onslaught of drought that occurred there on a regular basis, the picture would become complete.

So, most of the migrants whom I talked to attributed their decision to leave their previous homes directly or indirectly to their increasing vulnerability and entitlement failures. In times of food scarcity, when grains were sold in the markets they did not have the money to buy them. Asked whether they exhausted various coping strategies before they opted to move, the answers were variable. The great majority of the respondents who migrated to Gida Kiremu Woreda because of the existence of relatives in the destination area went there without exhausting the various coping strategies while others tried it partially, going to the extent of only borrowing grains and conducting seasonal migration before they made the final decision to move.

6.2. *Causes of Conflict and Displacement*

Ethnic and resource-based conflicts have played a role in shaping the political dynamics of almost all states in the Horn of Africa. In Ethiopia, the objectives of successive regimes to centralize the state and strengthen the power of the ruling class were facilitated by the marginalization and/or exclusion of minorities. Both the imperial regime and the military government opted for the process of modernization through the

---

[9] They were interviewed in their current living place, namely Jawi Resettlement Site, where they have been resettled since May 2001.

use of force. This aspect was succinctly put by Young, who stated that, "there was little scope in this process for the integration of the various ethnic groups, beyond the selective incorporation of individuals who accepted assimilations into the Amhara culture and society'' (1996:533). This historical legacy has left its imprint in present day Ethiopia. Different ethnic groups challenged the military dictatorship that ruled Ethiopia between 1974 and 1991 because it failed to reduce the power of the center and the regime was determined to crush those groups opposing its policies by force. After the collapse of the military government in 1991, the EPRDF-led government prepared a national conference to form a transitional government. Those that participated in the conference were dominantly ethnic-based political groups who took up arms and fought against the previous regimes for decades. In order to promote its own political agenda and those political groups that had been allied to it, the EPRDF-led government divided the country into various regions based on ethno-linguistic criteria. The regionalization has given a loophole for some ethnic parties and elites alike to use ethnic cards for vengeance and to fuel up ethnic grievances in one or the other parts of the country. As will be discussed shortly, the case study from Eastern Wollega has shown that environmental, political, socio-cultural, legal, and economic factors have incited inter-ethnic clashes resulting in the ousting of thousands of migrants.

Before dwelling upon the causes of conflict in the study area, it seems logical to discuss why conflicts occur in various areas in Ethiopia. One study (Ayele and Getachew, 2001) attempted to draw the conflict map of Ethiopia. By so doing, they have come up with dozens of flashpoints signalling internal conflicts in many parts of the country. The causes for the conflicts have been grouped into the following four categories by the authors: (a) *resource-based* (particularly land and water resources): e.g. Afar and Issa as of the early 1950s, Sidama ane Kembata after 1975, Afar & Amhara in N.Shewa from as far back as the 1920s, etc.; (b) *boundary issues*: Borena and Gari, Gedeo and Gujji, Gurra and Arsi, (c) *language*: Welayta and (d) *status*: Dire Dawa, Harari etc. Contrary to some newspaper reports, most, if not all, of these conflicts did not start after the EPRDF seized power in 1991 but rather rolled over from the imperial times through the Deg times with some of them becoming worse and others better. It will be misleading to sort multi-causal and multi-faceted conflicts into one or the other of the above boxes. Even in an apparently ethnic conflict, not all members of the particular ethnic groups show the same attitude and behaviour in the conflict.

Coming back to the case study proper, the lives of migrants-cum-settlers were disrupted when inter-ethnic conflicts flared up in Gida Kiremu Woreda first in March/April 2000 and later on in June 2000 (see to the chronology of events in Table 2). As stated by Deribssa (2004), the Amhara migrants were accused of establishing land claim, breaching contractual agreements entered into with the local population, which, the latter said, the migrants used as a cover to 'rob over land', 'abuse and misuse' of the forest resources, 'christening' place names by giving them Amharic names, carrying illegal firearms, cattle raiding, and demanding to create their own zone in the Woreda. As will be discussed shortly, almost all the accusations were contested by the migrants. In the words of the latter, political/ethnic factors, most particularly the agitations that were made against the migrant population by the local ethnic elites played a crucial role in triggering the conflict leading to their eventual displacement.

Table 2. Chronology of Events

Pre-1980	Spontaneous migration of Amharas from various zones of ANRS (Wello, South Gondar, East and West Gojam, Awi) into Eastern Wollega, most particularly into Gida Kiremu Woreda
Post-1980	Government-sponsored resettlement of thousands of Amhara peasants from ANRS (most particularly from Wello) to East Wollega
March/April 2000	First round fighting between the local population/administrators and the migrants-cum-settlers
June 2000	Second round fighting between the local population/administrators and the migrants-cum-settlers
November/December 2000	Third round of violent conflict that led to the displacement of 12,196 people or 3500 households

Source: Ethiopia: An Inventory of 28 Conflicts, Confidential Report, April 2003 (adapted)

The first conflict that took place in March/April 2000 was somehow contained, albeit temporarily, after the elders from both sides intervened to settle the problem. After a lull of about four to five months, the conflict resurfaced again in November/December 2000 claiming the lives of many people from both sides. A military intervention that followed resulted in the displacement of more than 12,000 Amhara cum-settlers or over 3500 households (refer to Tables 1 & 2). The violent clashes resulted in human casualties, the looting of cattle and grain, the burning of field crops and houses, the imprisonment of about 80 settlers, the burning of churches and the loss of property.

After the November/December 2000 clash, the settlers fled in panic to Burie, town located on the other side of the Abbay Valley in the southern flank of the Amhara Region with their meagre belongings and children. The ANRS Disaster Prevention and Preparedness Commission (DPPC) took the responsibility of accommodating the displaced population by putting up temporary plastic shelters in Burie. The Commission also took the responsibility of distributing relief handouts in the Burie camp up until their resettlement in Jawi/Metekel in May 2001. In the new resettlement site, about 265,281 hectares of land was cleared for homesteads and farm plot for the settlers. Settlers were given oxen for ploughing and food aid by the DPPC. All in all, the DPPC and the Ethiopian Ministry of Finance spent 1,002,707 Birr (US$ 117,966) for sheltering the displaced people in Burie and 3,269,077 Birr (US$ 384,597) for resettling them in Jawi between May 2001 and October 2002 (DPPC, November 2002).

The question now is what factors triggered the violent conflicts that led to the displacement of thousands of people? In what follows, the major causal factors that triggered the conflict as stated by some writers (e.g. Deribssa 2004), Woreda officials, the local population, the migrants, and the author's own judgement, will be highlighted under five categories, viz. environmental, political, socio-cultural, legal and economic.

(i)   **Environmental:** the local population complained that the migrants have indiscriminately cut trees to clear land for farming. They said that such acts affected their economic life and dietary diversifications; there were now few trees on which to hang beehives to produce honey. The migrants, in turn, have acknowledged the cutting of trees to clear the land for cultivation but said they were selective, leaving bigger trees untouched. It should at this juncture be affirmed that a sizeable proportion of the original forest cover has been lost due to the continuous process of migration and settlement in the study area.

(ii) **Political:** four factors that have influenced or triggered the conflict include: (a) a*gitation by the local elites against the presence of settlers in the Woreda.* This has been one of the major factors that fuelled tension, triggered violent clashes, eventually leading to the displacement of some of the settlers. The settlers consider all the actions taken against them as 'ethnic cleansing'. The author can also testify to the role of Woreda officials and cadres in politicising ethnicity and escalating the conflict. The officials showed partisanship, favouring their own kin. The Woreda officials, however, deny such allegation stating that not all Amharas living and working in the Woreda were targeted but only the 'outlaws'. How can our actions be considered as 'ethnic cleansing' they asked, when there are more than 18,000 Amhara settlers still living and working in different PAs and towns in Eastern Wollega zone? The existence of such a large number of Amhara settlers could be attributed not as claimed on legal grounds but rather on geographical location and mere chance factors. The great majority of the settlers live in the hitherto unoccupied low lying areas of the Woreda, away from the established villages inhabited by the local population. There were some settlers, who by chance have lived uphill interspersed with the local population. These have escaped the eviction and displacement that surfaced in eastern wollega in November/December 2000 while the lowlanders became easy targets and preys. (b) *The widely-rumoured allegation about the migrants demand to be considered as a special Amhara Zone within ONRS in a similar fashion to the Oromia Special Zone centred around Kemissie in ANRS.* As testified by the Eastern Wollega authorities, there was a misinterpretation of facts regarding this issue. The Amhara clergy and church goers alike created an ecclesiastical association in Eastern Wollega and named it as 'The Association of Guten Woreda Churches'. The *de facto* affixation of the word 'Woreda' to Guten, which did not exist on a *de jure* basis, coupled with the preparation of seals and stamps bearing that name without the permission of relevant regional offices led not only to the above-stated allegation but to the imprisonment of a number of priests and settlers. (c) The Woreda administration's decision to exclude all non-Oromos from PA leadership after 1991 acted as a point of dissatisfaction among the settlers. The exclusion was justified on the bases of language and ethnicity. Officially, Afan Oromo became the lingua franca and *kube* the script of the PAs.

(iii) **Socio-cultural:** some four factors that have a socio-cultural dimension could be included in this category: (a) t*he 'christening' or 'renaming' of places in the Woreda*: There is a consensus amongst my Oromo informants, farmers as well as authorities regarding the 'deliberate' change of place names from Oromiffaa into Amharic. Some examples of this include the change of 'Aaroo' into 'Haro Addis Alem'; 'Ajana' into 'Azana Selassie'; 'Dekkaa Jegi' into 'Shasho Ber'; and 'Bagin' into 'Bagin Mariam'. One of the justifications given by the settlers for the change or modification of place names has something to do with their inability to properly pronouncing local place names; the other is the identification of place names with newly established churches, which is the norm in their ancestral homes. (b) *Expansion of Ethiopian Orthodox Churches.* The settlers have established churches in almost all the villages they live in. The local people resented the migrants' decision to construct churches for it was made without consulting them. (c) The livelihood improvements that the migrants had attained in a short time (better life, control of markets and commerce) and the resultant jealousy that it kindled amongst the local population, and (d) *the chauvinistic attitude of the migrants towards the local population.* To the dismay of the local population, the historical bias of Amhara against the Oromos was by and large reflected in the study area. The migrants considered themselves as superior to the local population in terms of language and culture. This created a rift between the two populations.

(iv) **Legal:** included in here are three factors (a) *The breeching of contractual agreements over land by the settlers and the local population.* There have been many cases to breech of contract registered in PA and Woreda courts. There have been cases where some members of Oromo families sell land to the settlers but other claimants, mainly wives and grown-up children, refuse to recognize the deals and take the case to the courts in Gida Kiremu Woreda. The settlers complained that the courts have shown partisanship, favouring the local population, telling them to either make repayments or return the land to the original holders. Refusal to hand over the land back to the local population by the settlers resulted in armed clashes with the local police and militia. (b) *Refusal by the settlers to register their arms and surrender it to the police upon request.* The migrants complained that the action was selective, in the sense that the order was directed against them and not on the local population. The Amhara settlers rifle culture was further strengthened by the escalating cattle raiding as of mid-2000. (c) *The defiance by the post-1995 migrants to register in the PAs and Woreda offices* for fear that they would be expelled from the area as 'outlaws'.

(v) **Economic:** The sale of land to the ever-increasing migrant population has diminished the land available to the local farmers so much so that families have been unable to provide to give land to grown-up children.

7. Conclusions and Policy Implications

7.1 Conclusions

The study has tried to establish relationships between spontaneous migration, ethnicity, and resource conflict. The push factors that led people to migrate to other areas, in this particular case to Eastern Wollega, and the problems they have encountered in the host environment has also been noted. The degree to which resource conflict and social differentiation affect migration and population displacement in the context of resource scarcity and ethnicity has been discussed.

Of the many possible factors that forced the migrants in the study to leave their ancestral homes for good, about one-third of them are physical-environmental by nature. Included in these are recurrent drought, land and soil degradation, and declining soil fertility. The remaining push factors that have a socio-economic and political nature include social differentiation, weak traditional systems, greater exposure to risks, increasing vulnerability and entitlement failures. Similarly, some of the factors that attracted the migrants to Gida Kiremu Woreda are characteristically physical-environmental by nature. The availability of vacant or uninhabited fertile lands in the low lying areas of the Woreda that were avoided by the local population, and the suitability of the climatic condition for humans and livestock acted as centripetal forces in attracting the migrants to Gida Kiremu Woreda. The non-physical/environmental factors that pulled people towards the recipient Woreda include peer and kin pressure and the opening up of the Burie-Nekemt road.

Environmental, political, socio-economic, legal and economic factors triggered inter-ethnic conflicts in the study area eventually leading to population displacement. The excessive cutting down of trees (deforestation) to clear the land for farming has been considered by the local population and officials as a more decisive factor than others in inciting inter-ethnic conflict in Gida Kiremu Woreda. The non-environmental factors that provoked the clashes include disagreements over land rent contractual agreements between the local population and the settlers, agitation and hate campaigns waged against the settlers by the local ethnic elites, refusal to register and/or surrender firearms, rebuffing the Woreda administration's order to enlist new arrivals (migrants) in PA offices, deliberate exclusion of migrants from PA leadership, expansion of churches by the settlers, renaming or christening of place names in Amharic, envy and jealousy by the local population and local officials at the economic success of some of the migrant population, and the chauvinistic attitude of migrant-cum-settlers towards the Oromo population.

Owing to firstly, the presence of thousands of Amhara settlers still working and living in Gida Kiremu Woreda and secondly, the undiminished influx of migrants and returnees to the Woreda, the Federal Government and concerned authorities need to devise Conflict Prevention, Management and Resolution (CPMR) mechanisms that could possibly control the flaring up of yet another conflict. In what follows the policy implications of the study in the context of CPMR mechanisms will be pinpointed.

7.2 Policy Implications

Narrating the tragic and unfortunate situation and identifying the root causes of migration and conflict are not ends by themselves. One needs to learn from the past and devise mechanisms to deter the re-emergence of conflicts and the escalations thereof in any conflict-ridden area in Ethiopia. Given the facts that (a) there are still about 18,000 Amharas living and working in Gida Kiremu Woreda, (b) migration towards Wollega still continues and (c) some of the IDPs from Jawi in Metekel are returning to Wollega, there are possibilities for conflicts to flare up again. What lessons can be learnt from the past conflict? What can be done to avert such kinds of situations in the future? About six points can be raised to address these issues:

- In order to genuinely address the underlying causes of conflicts and to provide a framework by which all the parties to the conflict can and will voice their divergent interests so as to reach compromises, we need to break the long-held taboo of covering up conflicts in Ethiopia. One elderly informant, who had migrated to Gida Kiremu Woreda back in 1972 and still lives in Kiremu rural town, stated that even the government itself and state-run media maintained a conspiracy of silence when thousands of people were displaced and hundreds died during the conflict that flared up in the Woreda in 2000. Ironically, stated the same informant, we repeatedly hear from the same media reports on relatively minor events such as car crashes claiming the lives of a couple of people.
- Although it may be difficult to stop the migratory flow of people, government policies should be designed that could help in (i) averting and controlling the adverse effects of inappropriate development strategies, (ii) promoting socio-economic and political policies that expand the base for sustainable development through improvements in the entitlements of people (iii) making people less vulnerable towards natural hazards, ensuring their access to natural resources and developing other alternatives than farming. These measures can keep people in their ancestral homes.
- Mechanisms should be sought by which the federal government's devolution of power to nine regional states based on ethnic lines (ethnic-based decentralization) should no more be used to promote political agendas by politicising ethnicity. Ethnic elites in different regional states of should be advised not to employ ethnic identities to mobilize support and to incite communal hatred for political ends by stereotyping and defaming other groups.
- As successful conflict resolution mechanisms between the Afars and Issas in 1998 and between Amharas, Argobas, Oromos and Afars in North Shewa zone have clearly demonstrated, we need to enhance and strengthen traditional conflict management systems (e.g. Shimgilna) by establishing Joint Peace Committees (JPCs) that would be composed of all stake-holders, including the warring parties. Prior to the eruption of the conflict in November/December 2000, attempts were made to resolve the differences between the settlers and the local population by establishing a JPC that was made up of seven elders from both sides. The Committee's activities were curtailed when acts such as cattle raiding overtook events resulting in armed clashes.

- Inter-regional cooperation (e.g. between ANRS and ONRS in the present case is also required to promote reconciliation between the warring parties based on good faith. By way of cultivating those traditional mechanisms, community peace building groups should be created that could possibly develop into cross-border community peace building institutions. Such an attempt was also made in Eastern Wollega in the summer of 2000, about three months prior to the eruption of the conflict, when ANRS representatives from Gojjam, Wollo and Gondar were sent to the troubled region to jointly sort out the problems with their counterparts in Eastern Wollega. The ANRS representatives stayed there for a couple of months talking not only with authorities in Eastern Wollega but with the settlers as well. The representative of East Gojam zone, Ato Chane Shimekaw, informed the author about their success in disarming the settlers by which no less than 200 settlers voluntarily surrendered their firearms in a short time. The same informant stated that splits occurred between them and Eastern Wollega officials on issues related to exposing criminals on both sides. The former volunteered to do so while the latter refused. This, in turn, created cleavages between the two parties frustrating the reconciliation attempt. After the failure of the rapprochement, the ANRS representatives left Eastern Wollega and submitted a report indicating the inevitability of conflict.

- As stated by Hauss (2001), the clash of interests may not be compromised unless one of the parties gives up and/or steps back from specific demands that are antagonistic to the other. The Federal Government of Ethiopia should create a common ground by which competing groups should come to terms by discussing their differences. Revealing the truth, indicting the guilty, building social capital, (because internal strife can obviously destroy social ties in the society), enhancing indigenous and traditional ways of conflict resolution mechanisms can help in bringing peace and stability in conflict-ridden areas.

References

Adedeje, Adebayo (ed.) [1999]. Comprehending and Mastering of African Conflicts: The Search for Sustainable Peace and Good Governance. London: Zed Books.

Assefa Tolera (1999). Ethnic Integration and Conflict: The Case of Indigenous Oromo and Amhara Settlers in aaroo Addis Alem, Kiramu Area, Northeastern Wollega. Department of Sociology and Social Anthropology, Dissertation Series No. 5.

Ayele G.mariam and Getachew Kassa (October 2001). Conflict Prevention, Management, & Resolution: Capacity Assessment Study for the IGAD Sub-region. University of Leeds, Center for Development Studies.

Blaikie, P et al (1994). At Risk: Natural hazards, people's vulnerability, and disasters. London: Routledge.

Bohle, H.G. (2001), "Vulnerability and Criticality: Perspectives form Social Geography" in IHDP Update 2, pp. 1-4.

Bohle, H.G. and Watts, M.J. (1993) "The Space of Vulnerability: The Causal Structure of Hunger and Famine", *Progress in Human Geography* Vol.13, No. 1, pp.43-67.

Bohle, H.G. et al (1994), "Climate Change and Social Vulnerability: Toward a Sociology and Geography of Food Insecurity", *Global Environmental Change*, Vol. 4, No. 1, pp. 37-48.

Cannon, T. (1994), "Vulnerability Analysis and Natural Disasters", in Varley, A. (ed.). Disaters, Development and Environment. Chichester: Wiley and Sons Ltd.

Carney, D (ed.) [1988]. Sustainable Rural Livelihoods. London: Department of International Development (DfID).

Chambers, R. (1989), "Vulnerability, Coping and Policy", *IDS Bulletin* Vol. 20, No. 2, pp. 1-7.

Corbett, J. (1988), "Famine and Household Coping Strategies", *World Development*, Vol. 16, No. 9. pp. 1099-1112.

Deribssa Abate (2004). Decentralization and the Management of Ethnic Conflict in Ethiopia: handling ethnic conflicts in the peripheral areas of ONRS, the case of Amhara vs. Oromo ethnic conflict in East Wollega zone, OSSREA Research Report No.

Dessalegn Rahmato (1991). Famine and Survival Strategies: A Case Study from Northeast Ethiopia. Uppsala: Nordsika Afrikainstitutet.

De Waal, Alex (1988), "Famine Early Warning Systems and the use of Socio-economic Data", *Disasters*, Vol. 12, No. 1, pp. 81-91.

Disaster Prevention and Preparedness Commission, ANRS (2002). Jawi Resettlement Project Report. Bahir Dar.

Ethiopia: An Inventory of 28 Conflicts, Confidential Report, April 2003.

Gebre Yntiso (March 2002), "Differential Reestablishment of Voluntary and Involuntary Migrants: The Case of Metekel Settlers", *Africa Study Monograph*, 23, 1, 31-46.

Hauss, Charles (2001) International Conflict Resolution. London: Continuum.

Hewitt, K (1983). Interpretations of calamity from the viewpoint of Human Ecology. Boston.

Hutchison, Robert A (ed.) [1991]. Fighting for Survival: Insecurity, People and the Environment in the Horn of Africa. Gland: The World Conservation Union (IUCN) Publication.

Lonergan, S. (1988), "The Role of Environmental Degradation in Population Dispalcement", Research Report 12, Victoria, BC.

Markakis, John (1998). <u>Resource Conflict in the Horn of Africa</u>. London: The Cromwell Press Ltd.

Maybury-Lewis, David (ed.) [1997]. <u>Indigenous Peoples, Ethnic Groups, and the State</u>. Boston: Allyn and Bacon.

Mesfin Wolde Mariam (1984). <u>Rural Vulnerability to Famine in Ethiopia: 1958-1977</u>. New Delhi: Vikas Publishing House Pvt. Ltd.

Sen, A.K. (1981). <u>Poverty and Famines: An Essay on Entitlements and Deprivation</u>. Oxford: Clarendon Press.

Suhrke, Astri (1994), "Environmental Refugees and Social Conflict", in Baechler (ed.). <u>Umweltfluechtlinge: Das Konfliktpotential von Morgen</u>?. Muenster: Lit Verlag.

Wenzel, H.-J. (2002), "Umweltfluechtlinge oder Umweltmigranten? Umweltdegradation, Verwundbarkeit und Migration/Flucht im subsaharischen Afrika", IMIS Schriften, Bd.11, S. 287-311.

Young, John (1998), "Regionalism and Democracy in Ethiopia", *Third World Quarterly*, Vol. 19, No. 2. pp. 191-204, 1998.

Persons Interviewed

1. Ato Dereje Getahun, Deputy Administrator of Gida Kiremu Woreda, June 1, 2003, Ayana.

2. Ato Assefa Feyissa, Administrator of Gida Kiremu Woreda at the time of the conflict, May 31, 2003, Ayana.

3. Ato Yared Keneaa, Current Administrator of Gida Kiremu Woreda, June 1, 2003, Ayana

4. Ato Wuletaw H/Mariam, Director of the Organization for Rehabilitation and Development of Amhara (ORDA), May 27, 2003, Bahir Dar

5. Ato Melakm Endale, Group Leader of ANRS's DPPC Early Warning Department and Team Member of the Jawi Resettlement Project, Enjbara, May 29, 2003.

6. Ato Getahun Atnafu, ANRS/DPPC, Early Warning Department Expert, May 26, 2003, Bahir Dar.

FSS Publications List

FSS Periodical

Medrek, now renamed BULLETIN (Quarterly since 1998. English and Amharic)

FSS Discussion Papers

No. 1 *Water Resource Development in Ethiopia: Issues of Sustainability and Participation.* Dessalegn Rahmato. June 1999

No. 2 *The City of Addis Ababa: Policy Options for the Governance and Management of a City with Multiple Identity.* Meheret Ayenew. December 1999

No. 3 *Listening to the Poor: A Study Based on Selected Rural and Urban Sites in Ethiopia.* Aklilu Kidanu and Dessalegn Rahmato. May 2000

No. 4 *Small-Scale Irrigation and Household Food Security. A Case Study from Central Ethiopia.* Fuad Adem. February 2001

No. 5 *Land Redistribution and Female-Headed Households.* By Yigremew Adal. November 2001

No. 6 *Environmental Impact of Development Policies in Peripheral Areas: The Case of Metekel, Northwest Ethiopia.* Wolde-Selassie Abbute. Forthcoming, 2001

No. 7 *The Environmental Impact of Small-scale Irrigation: A Case Study.* Fuad Adem. Forthcoming, 2001

No. 8 *Livelihood Insecurity Among Urban Households in Ethiopia.* Dessalegn Rahmato and Aklilu Kidanu. October 2002

No. 9 *Rural Poverty in Ethiopia: Household Case Studies from North Shewa.* Yared Amare. December 2002

No.10 *Rural Lands in Ethiopia: Issues, Evidences and Policy Response.* Tesfaye Teklu. May 2003

No.11 *Resettlement in Ethiopia: The Tragedy of Population Relocation in the 1980s.* Dessalegn Rahmato. June 2003

No.12 *Searching for Tenure Security? The Land System and New Policy Initiatives in Ethiopia.* Dessalegn Rahmato. August 2004.

FSS Monograph Series

No. 1 *Survey of the Private Press in Ethiopia: 1991-1999.* Shimelis Bonsa. 2000

No. 2 *Environmental Change and State Policy in Ethiopia: Lessons from Past Experience.* Dessalegn Rahmato. 2001

No. 3 *Democratic Assistance to Post-Conflict Ethiopia: Impact and Limitations.* Dessalegn Rahmato and Meheret Ayenew. 2004

FSS Conference Proceedings

1. *Issues in Rural Development. Proceedings of the Inaugural Workshop of the Forum for Social Studies, 18 September 1998.* Edited by Zenebework Taddesse. 2000

2. *Development and Public Access to Information in Ethiopia.* Edited by Zenebework Tadesse. 2000

3. *Environment and Development in Ethiopia.* Edited by Zenebework Tadesse. 2001

4. *Food Security and Sustainable Livelihoods in Ethiopia.* Edited by Yared Amare. 2001

5. *Natural Resource Management in Ethiopia.* Edited by Alula Pankhurst. 2001

6. *Poverty and Poverty Policy in Ethiopia.* Special issue containing the papers of FSS' final conference on poverty held on 8 March 2002

Consultation Papers on Poverty

No. 1 *The Social Dimensions of Poverty.* Papers by Minas Hiruy, Abebe Kebede, and Zenebework Tadesse. Edited by Meheret Ayenew. June 2001

No. 2 *NGOs and Poverty Reduction.* Papers by Fassil W. Mariam, Abowork Haile, Berhanu Geleto, and Jemal Ahmed. Edited by Meheret Ayenew. July 2001

No. 3 *Civil Society Groups and Poverty Reduction.* Papers by Abonesh H. Mariam, Zena Berhanu, and Zewdie Shitie. Edited by Meheret Ayenew. August 2001

No. 4 *Listening to the Poor.* Oral Presentation by Gizachew Haile, Senait Zenawi, Sisay Gessesse and Martha Tadesse. In Amharic. Edited by Meheret Ayenew. November 2001

No. 5 *The Private Sector and Poverty Reduction [Amharic].* Papers by Teshome Kebede, Mullu Solomon and Hailemeskel Abebe. Edited by Meheret Ayenew, November 2001

No. 6 *Government, Donors and Poverty Reduction.* Papers by H.E. Ato Mekonnen Manyazewal, William James Smith and Jeroen Verheul. Edited by Meheret Ayenew, February 2002.

No. 7 *Poverty and Poverty Policy in Ethiopia.* Edited by Meheret Ayenew, 2002

Books

1. *Ethiopia: The Challenge of Democracy from Below.* Edited by Bahru Zewde and Siegfried Pausewang. Nordic African Institute, Uppsala and the Forum for Social Studies, Addis Ababa. 2002

Special Publications

- *Thematic Briefings on Natural Resource Management. Enlarged Edition.* Edited by Alula Pankhurst. Produced jointly by the Forum for Social Studies and the University of Sussex. January 2001

New Series

- ### Gender Policy Dialogue Series

No. 1 *Gender and Economic Policy.* Edited by Zenebework Tadesse. March 2003
No. 2 *Gender and Poverty (Amharic).* Edited by Zenebework Tadesse. March 2003
No. 3 *Gender and Social Development in Ethiopia.* (Forthcoming).
No. 4 *Gender Policy Dialogue in Oromiya Region.* Edited by Eshetu Bekele. September 2003

- ### Consultation Papers on Environment

No. 1 *Environment and Environmental Change in Ethiopia.* Edited by Gedion Asfaw. Consultation Papers on Environment. March 2003

No. 2 *Environment, Poverty and Gender.* Edited by Gedion Asfaw. Consultation Papers on Environment. May 2003

No. 3 *Environmental Conflict.* Edited by Gedion Asfaw. Consultation Papers on Environment. July 2003

No. 4 *Economic Development and its Environmental Impact.* Edited by Gedion Asfaw. Consultation Papers on Environment. August 2003

No. 5 *Government and Environmental Policy.* Consultation Papers on Environment. January 2004

No. 6 *የግልና የጋራ ጥረት ለአካባቢ ሕይወት መሻሻል (የስሜን ሸዋ ገበሬዎች ተሞክሮ)* Consultation Papers on Environment. May 2004

No. 7 *Promotion of Indigenous Trees and Biodiversity Conservation.* Consultation Papers on Environment. June 2004

- ### FSS Studies on Poverty

No. 1 *Some Aspects of Poverty in Ethiopia: Three Selected Papers.* Papers by Dessalegn Rahmato, Meheret Ayenew and Aklilu Kidanu. Edited by Dessalegn Rahmato. March 2003.

No. 2 *Faces of Poverty: Life in Gäta, Wälo.* By Harald Aspen. June 2003.

No. 3 *Destitution in Rural Ethiopia.* By Yared Amare. August 2003

No. 4 *Environment, Poverty and Conflict.* Tesfaye Teklu and Tesfaye Tafesse. October 2004

Educational Materials

- Environmental Posters
- Environmental Video Films
  - የከተማችን አካባቢ ከየት ወዴት?
  - ውጥንቅጥ፤ ድህነትና የአካባቢያችን መጎሳቆል

www.ingramcontent.com/pod-product-compliance
Lightning Source LLC
Chambersburg PA
CBHW080844270326
41929CB00016B/2918